Made in United States
Orlando, FL
07 November 2026

90337530R00057